GEOTHERMAL TREASURES

—

MĀORI LIVING WITH HEAT AND STEAM

CONTENTS

ĪHENGA

THE DISCOVERY OF ROTORUA

Tamatekapua was the captain of Te Arawa waka (canoe) that journeyed from Hawaiki. His grandson, Īhenga, is credited with discovering Rotorua and naming many of its places. For example, the full name of Lake Rotoiti is Te Rotoiti-kitea-e-Īhenga-i-Ariki-ai-Kahu, the narrow lake seen by Īhenga, and the full name of Lake Rotorua is Te Rotorua-nui-a-Kahumatamomoe – the second large lake of Kahumatamomoe, who was Īhenga's uncle.

People who claim descent from the Te Arawa waka had originally settled on the coast at Maketū, but after Īhenga returned with news of his discoveries inland, some of the tribe migrated to the Rotorua and Taupō areas. Īhenga named Ōhinemutu on the shores of Lake Rotorua for his people. They found the thermal activity suitable for their cooking, bathing and heating needs, and Māori living at Ōhinemutu and Whakarewarewa still use it in these ways today.

GEOTHERMAL TREASURES

As you drive or fly into Rotorua, you see steam rising across the district from numerous areas of thermal activity, including Tikitere in the northeast, Kuirau Park in the west and Whakarewarewa and Waiotapu to the south. The effect is stunning, especially on cool, sunny mornings, when a glassy crater lake rests in the foreground. Pockets of steam also rise up across the city from Ngapuna and Sulphur Point and, most days, the distinctive smell of sulphur permeates the air.

For Māori, geothermal heat is a taonga (treasure) of great historical, cultural, spiritual and economic importance. This is especially so for the North Island iwi (tribes) of Te Arawa (Rotorua), Tūwharetoa (Taupō) and Mataatua (Bay of Plenty) living on the volcanic plateau of the central North Island. For Māori, ngā waiariki (the thermal springs) are a highly prized ancestral legacy.

The earth's energy was central to the traditional Māori way of life. In geothermal areas, this is still true today, where local Māori use it to cook, heat, bathe, heal and purify. Geothermal energy is also used to generate power for the national grid, with some local communities benefitting financially from the use of their land and resources.

The chance to experience natural geothermal phenomena has long been a drawcard for tourists. Visitors are awestruck by the beauty and magic of hot water gushing from beneath the earth.

There are few places in the world where traditional indigenous stories and geothermal science merge as strongly as they do in this region.

The hot pools that have made Rotorua famous come in all shapes, sizes and temperatures, ranging from tepid to boiling hot. The therapeutic benefits of these mineral waters have long been known. Since 1908, the Bath House at the Rotorua Museum has been visited by thousands of people from around the world, all looking for relief, particularly from ailments such as arthritis.

Today, people wanting treatment in the healing waters still flock to the city to bathe in the steaming mineral pools. Other visitors enjoy the many thermal tourist sights the city has to offer. Pools can be found in family homes, on private land in the middle of farms, as well as at different scenic attractions throughout Rotorua and surrounding districts. Some have been harnessed as commercial enterprises, while others are freely accessible to the public.

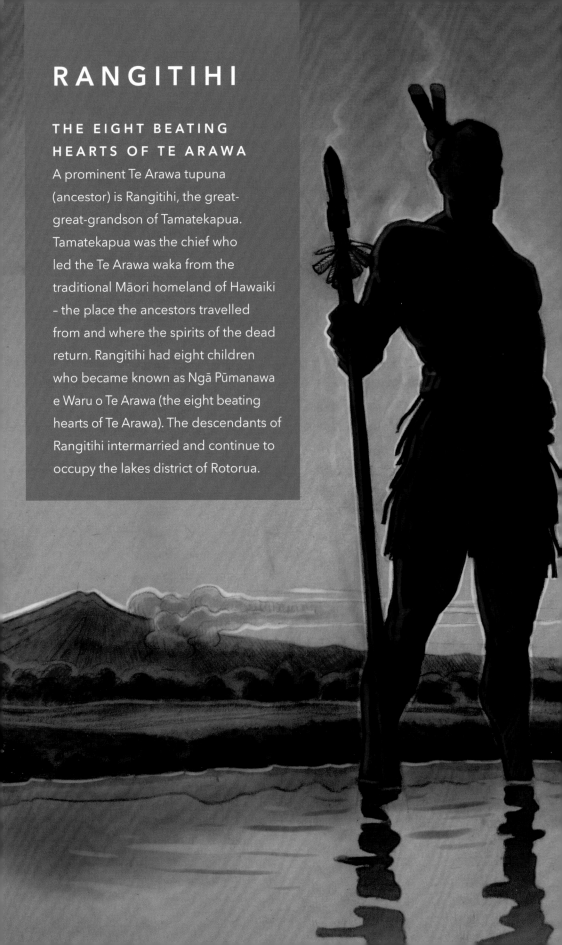

RANGITIHI

THE EIGHT BEATING HEARTS OF TE ARAWA

A prominent Te Arawa tupuna (ancestor) is Rangitihi, the great-great-grandson of Tamatekapua. Tamatekapua was the chief who led the Te Arawa waka from the traditional Māori homeland of Hawaiki – the place the ancestors travelled from and where the spirits of the dead return. Rangitihi had eight children who became known as Ngā Pūmanawa e Waru o Te Arawa (the eight beating hearts of Te Arawa). The descendants of Rangitihi intermarried and continue to occupy the lakes district of Rotorua.

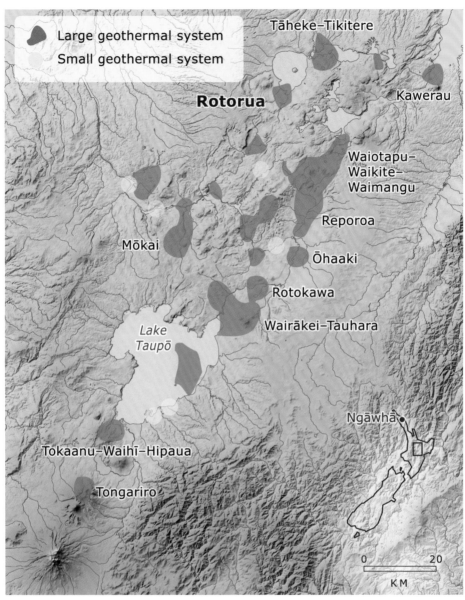

Large geothermal system
Small geothermal system

Tāheke–Tikitere

Rotorua

Kawerau

Waiotapu–
Waikite–
Waimangu

Reporoa

Mōkai

Ōhaaki

Rotokawa

Wairākei–Tauhara

Lake
Taupō

Ngāwhā

Tokaanu–Waihī–Hipaua

Tongariro

0   20
KM

Te Ara, the Encyclopedia of New Zealand http://www.teara.govt.nz/ © Crown Copyright 2005 – 2014 Manatū Taonga,
The Ministry for Culture and Heritage, New Zealand

This map shows the active geothermal area of the central North Island.
The area extends from Whakaari (White Island) in the Bay of Plenty (not
pictured) to Rotorua, a city within a caldera, south to Lake Taupō, another
flooded caldera, and on to the volcanoes of the Tongariro National Park.

Within this area, seventeen major geothermal fields contain 80 percent
of New Zealand's geothermal systems. Each field has unique surface features,
including hot pools, mud pools, geysers, fumaroles and silica deposits.

THE GEOTHERMAL LANDSCAPE

The volcanic and geothermal lands of the central North Island have strong spiritual and cultural significance for Māori, and particularly for the iwi of that area – Te Arawa, Tūwharetoa and Mataatua.

A number of deities are enshrined in traditional stories and legends associated with volcanic action, earthquakes and particular geothermal places or events.

In traditional Māori stories, Te Whakarewarewa Valley in Rotorua was one of the places where the deities of fire, Te Pupu and Te Hoata, emerged from the earth as they tried to bring the warmth of Hawaiki to Aotearoa. At the time Ngātoroirangi, the famous Te Arawa ancestor, was exploring the volcanic region, and he and his companion, Ngāuruhoe, were almost freezing to death in a snowstorm on Mount Tongariro. Te Pupu and Te Hoata surfaced in several places from their subterranean journey creating the geothermal activity that remains today at Whakaari off the coast of the Bay of Plenty as well as at Tikitere, Whakarewarewa, Waimangu, Waiotapu, Orākei Korako, Wairākei, Tokaanu and in the mountains Tongariro and Ruapehu.

RŪAUMOKO

GOD OF EARTHQUAKES AND VOLCANIC FIRE

In the Māori creation story, Ranginui (the sky father) and Papatūānuku (the earth mother) were separated by their offspring. Rūaumoko was their youngest, a baby, and his brothers gifted him ahi tipua (volcanic or supernatural fire) to provide warmth and comfort to their mother. Thus, Rūaumoko became the god of earthquakes and volcanic fire.

Māori describe fire in three ways: ahi māori (natural fire), ahi tipua (volcanic or supernatural fire) and ahi o te atua (fire of the gods).

MAHUIKA AND MĀUI

BRINGING FIRE TO THE WORLD

It was from Mahuika, the goddess of ahi māori, that the mischievous demi-god Māui obtained the secret of making fire. He tricked her into giving him the ahi māori from her fingernails and toenails. Mahuika is also a descendant of Te Rā (the sun), snared by Māui to slow its journey and lengthen the daylight hours. Famous for his exploits, Māui is also credited with bringing heat from Hawaiki, when he fished up the North Island, which is named Te Ika a Māui (The Fish of Māui).

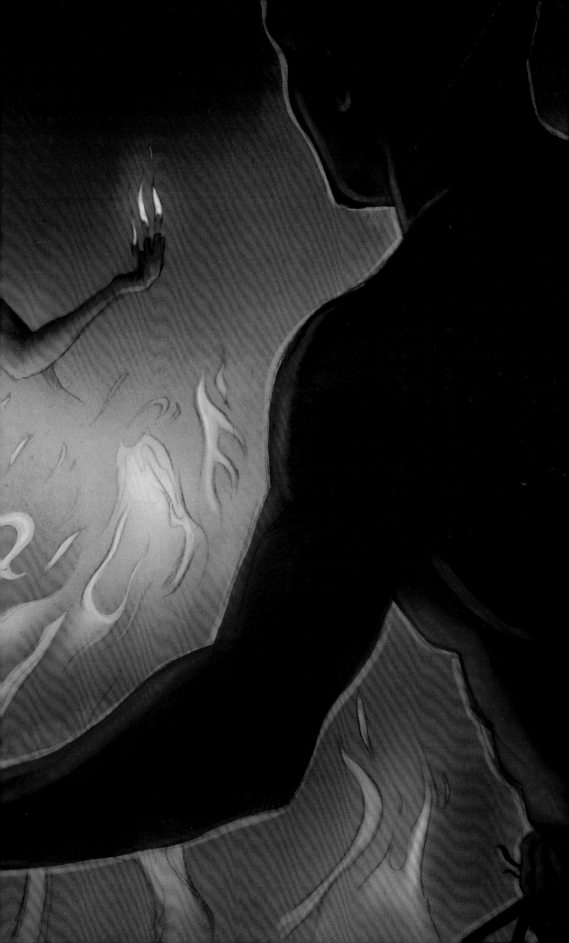

ROTORUA

A large volcanic crater, called a caldera, formed about 240,000 years ago in the central North Island. Within this caldera, both the city of Rotorua and Lake Rotorua lie today. Throughout the Rotorua basin, more than 600 geothermal features exist because of residual heat from the caldera eruptions that formed the volcanoes of Rotorua, Tarawera and Haroharo.

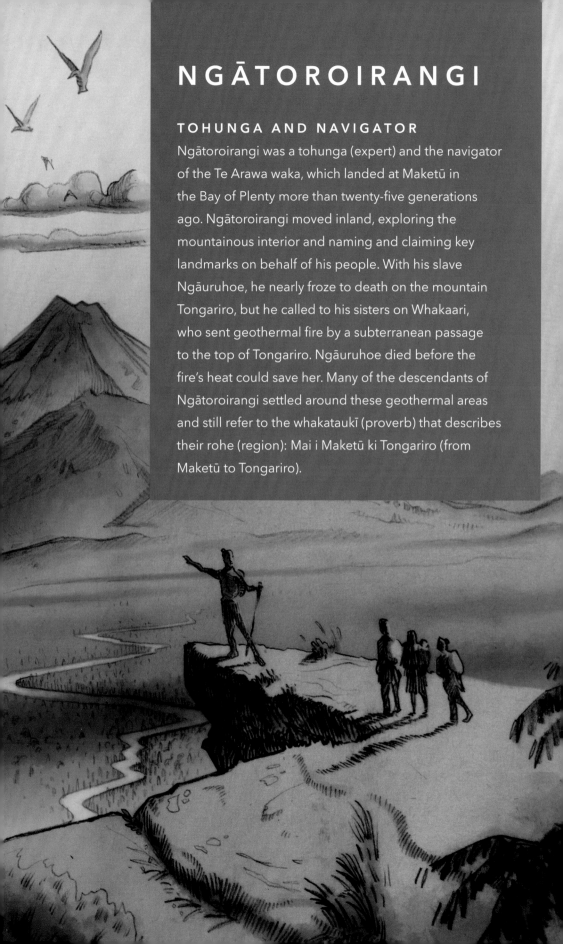

NGĀTOROIRANGI

TOHUNGA AND NAVIGATOR

Ngātoroirangi was a tohunga (expert) and the navigator of the Te Arawa waka, which landed at Maketū in the Bay of Plenty more than twenty-five generations ago. Ngātoroirangi moved inland, exploring the mountainous interior and naming and claiming key landmarks on behalf of his people. With his slave Ngāuruhoe, he nearly froze to death on the mountain Tongariro, but he called to his sisters on Whakaari, who sent geothermal fire by a subterranean passage to the top of Tongariro. Ngāuruhoe died before the fire's heat could save her. Many of the descendants of Ngātoroirangi settled around these geothermal areas and still refer to the whakataukī (proverb) that describes their rohe (region): Mai i Maketū ki Tongariro (from Maketū to Tongariro).

ME TĪMATA - TELLING A STORY THROUGH A SONG POEM

Me Tīmata (I begin) is a pātere (song poem), usually sung or chanted to a rhythmic beat.

It was written by Mauriora Kingi, a tohunga from Te Arawa and Tainui iwi and details the geothermal history and sites of Te Whakarewarewa Valley. Mauriora is a recognised advisor on Māori protocol and an authority on the stories of this world-famous geothermal reserve.

ME TĪMATA, PART 1

Me tīmata i a Ngātoroirangi
I pikitia ake ai i te tihi o Tongariro
Ka kinongia tana kiri i te mātao
Ka auē atu rā ki ōna tuāhine
Kia haria mai te ahi i te wā kāinga
Kia ngata rā te hiahia
Ka haruru mai te whenua ao te whenua pō
Ka tatū mai ki tana taha
Mahue atu i a rāua te wera me te ahi
I te tahatā ka mātotoru, ka māraratia
Noho iho rā te mamaoa
Ki te Whakarewarewa
Te tūnga o te ope taua a Wahiao e!

I begin my journey with Ngātoroirangi
Who climbed the peak of Tongariro
Suddenly his skin became very cold
Crying to his sisters
To bring him the fire from Hawaiki
To fulfil his desires
Both the underworld and the land rumbled
Soon arriving at his side
But leaving behind them traces of the journey being heat and fire
Where they rested and it became thick and spread
Leaving the steam
At Whakarewarewa
Where the uprising war party of Wahiao resided!

HINEMOA AND TŪTĀNEKAI

THE GREAT LOVE STORY

Mokoia Island in Lake Rotorua is the historical location of one of New Zealand's greatest love stories.

Hinemoa was the beautiful daughter of a high-ranking chief from Ōwhata, on the eastern shores of Lake Rotorua. She had many suitors and was expected to marry a man chosen by her family. Instead, she fell in love with Tūtānekai, who lived on Mokoia Island. Every evening, Hinemoa listened to Tūtānekai playing his bone flute from across the lake. Her family, however, prohibited them from marrying and beached all the waka so she could not join him on the island. In desperation, Hinemoa decided to swim to Mokoia.

She tied three calabashes to each arm and, guided by the sound of her beloved's music, managed to reach the island. The place she set out from is now called Hinemoa Point. Tūtānekai found his love recuperating from her epic swim in a hot pool on Mokoia, known today as Waikimihia or Hinemoa's Pool. Many carvings by Te Arawa artists feature Hinemoa and Tūtānekai, often depicting them with calabashes and a flute.

TE WHAKAREWAREWA VALLEY

Māori people have lived in the Rotorua area for more than twenty-five generations, enjoying the warmth and therapeutic benefits of its thermal environment. Today, the village of Whakarewarewa, in Te Whakarewarewa Valley, is a popular tourist attraction, as it has been since the 1880s.

Three hapū (sub-tribes) of Te Arawa have occupied Whakarewarewa at different times throughout its history. They are Tūhourangi, Ngāti Wahiao and Ngāti Whakaue.

As Māori settled in the area, they built pā (fortresses) throughout the landscape. Within the valley itself was the impenetrable pā of Te Puia (the volcano, geyser or hot spring) whose original occupants, around 1300 AD, were a people known as the dragon slayers. The pā was strategically built beneath the cliffs of the Pohaturoa mountains and surrounded by a natural moat of hot pools.

More than 200 years ago, a war party preparing to do battle with a neighbouring iwi was inspired by its leader, Wahiao, to launch into a ferocious haka (posture dance).

The performance was so spectacular that the local people named the area Te Whakarewarewatanga o Te Ope Taua a Wahiao (the gathering place for the war parties of Wahiao). Over the years, that name has been shortened to Te Whakarewarewa, Whakarewarewa or simply Whaka.

Te Whakarewarewa Valley is a globally unique environment of enormous geological significance. It is home to many unusual plants, animals and micro-organisms that survive in the hot alkaline and acidic waters and soil conditions. It is one of the most diverse collections of surface geothermal features in Aotearoa. These include mud pools, acid-sulphate pools, weakly enriched chloride pools, boiling springs and New Zealand's largest surviving collection of geysers. The Pohutu Geyser, one of fifty-five geysers in Te Whakarewarewa Valley, spouts to a height of 30 metres (100 feet) a number of times each day. Māori view themselves as kaitiaki (guardians) of these geothermal wonders, ensuring their conservation and preservation for future generations.

HATUPATU

ESCAPING THE BIRD WOMAN

A celebrated hero of Te Arawa is Hatupatu, whose family settled on the sacred Mokoia Island. He had many adventures, but his most famous feat was escaping from the fearsome bird-woman Kurangaituku, who lived in a cave near Rotorua. Hatupatu had been killed on a bird-hunting expedition, but was brought back to life through karakia (prayer). On his way home, though, he was captured and imprisoned by Kurangaituku. Hatupatu managed to escape, with Kurangaituku in full pursuit. On reaching the hot pools and geysers of Whakarewarewa, Hatupatu leaped over them to safety. Kurangaituku tried to wade after him, but she perished in the boiling mud.

ME TĪMATA, PART 2

Ka tū waewae ahau ki runga
i a Te Pākira
Kia mārama te titiro ki Whakatūtū
I mate ai a Kurangaituku i a Hatupatu
Kia kite au a Roto-ā-Tamaheke
Te rerenga waikaukau o ngā tūpuna
mātua
Me huri whakarunga ki Papawharanui
Kia toro taku ringa ki Parekōhoru
Te ohākītanga o te iwi e!

Now I stand with my two feet on
Te Pākira
Seeing clearly Whakatūtū, the hot
springs
Where Kurangaituku died whilst
chasing Hatupatu
Gazing over to see Roto-ā-Tamaheke
The running waters where our
ancestors bathed
I then turn up to Papawharanui
Reaching out my hand to Parekōhoru
The main source of the people!

THE 1886 TARAWERA ERUPTION

From as early as the 1870s, visitors from around the world came to the region to view the magical Pink and White Terraces (Otukapuarangi and Te Tarata) on the shores of Lake Rotomahana. The silica terraces – built up over 600 or 700 years of geothermal action – were called the eighth wonder of the world.

But the landscape changed forever on 10 June 1886, when the Ruawāhia Dome on Mount Tarawera erupted, killing at least 106 people. An ash cloud filled the night sky. People at nearby Te Wairoa village (known as the Buried Village) were woken by violent earthquakes as craters were forced open, spewing out ash, mud and steam.

The rumbling of the eruption was reportedly heard in Nelson, in the South Island, and there were earthquakes throughout the North Island. The famous terraces vanished beneath Lake Rotomahana, and a deep crater formed where the terraces once stood. Lake Rotomahana became much larger and new geothermal features emerged, including the largest geyser in the world – the Waimangu Geyser – which had a brief but spectacular life from 1900 to 1904.

TAMA-O-HOI AND TARAWERA

After the arrival of the Te Arawa waka, the tohunga
Ngātoroirangi travelled inland. He met a spirit called
Tama-o-hoi at Mount Tarawera. Tama-o-hoi claimed the
tohunga was trespassing on his land and attempted to
use sorcery to destroy him. However, Ngātoroirangi cast a
superior spell against his opponent, including by stamping
his foot on the summit of Mount Tarawera, causing a
chasm to open. Ngātoroirangi threw Tama-o-hoi into the
mountain. The 1886 eruption of Tarawera was blamed by
some on Tama-o-hoi. They said he was furious at having
been imprisoned in the mountain for such a long time
and that he got his revenge by causing this catastrophe.

ME TĪMATA, PART 3

Kia hīkoi au ki Korotiotio

Kei tua mai ko Te Waro-a-Ngāwai

Kei kō atu ko Te Wai-o-Nihotahi

Kia peke atu ki roto o Puārenga

Kia rere i te tai

Me mihi atu rā ki Ngā Mōkai-a-kōkō

E kōhuahua ra

Ko te Mitimiti, ko Waikōrua, ko
Parengātata, Ko Tūtaepoko

Then walking to Korotiotio

Also seeing Te Waro-a-Ngāwai on
one side

And Te Wai-o-Nihotahi on the other

Now jumping into Puārenga

To let the tide carry me

Until I greet Ngā Mōkai-a-kōkō

Boiling heartily is

Te Mitimiti, Waikōrua, Parengātata
And Tūtaepoko

WAIMANGU

The Waimangu Volcanic Valley lies on the opposite shore of Lake Rotomahana from Mount Tarawera. Along the tracks of this scenic reserve, visitors can witness amazing geothermal activity such as the colourful Warbrick Terrace, the bubbling Ngapuia o Te Papa Springs, the sizzling Waimangu Cauldron (Frying Pan Lake) and the deep turquoise water of Ruaumoko's Throat Crater. The Waimangu Cauldron is a hot lake covering four hectares, perhaps the largest hot spring in the world. Pink, white and greyish-red cliffs rise from the waters, caused by springs of natural silica disgorging from beneath the earth. Recently, thousands of gas vents have been discovered spurting from the lake bed.

Photograph: Waimangu Volcanic Valley

ME TĪMATA, PART 4

Me hoki au ki te awa
Kei kō mai ko Waihuka, ko Tikorangi
I wera mai te moana i a Te Korokoro
Ko te hoa ko Kererū
Ko te wai ōranga o rātau
Ko Waikorohīhī, ko te Māhanga,
ko Komutumutu
Ka heke ai te roke ko Waiparu
Kia whiti atu ki Puapua
E tū whakahirahira mai rā ko Pōhutu
Te whakamīharotanga o te ao!

Then returning back to the (Puārenga)
stream
And at my side is Waihuka, Tikorangi
Heating the stream is Korokoro
And Kererū
Stopping to bathe in the healing pools
Of Waikorohīhī, Māhanga and
Komutumutu
Also watching the waste flow into
Waiparu
I reach out to Puapua and I see
Standing proudly, Pōhutu
The admiration of the world!

ME TĪMATA, PART 5

E pōhiri mai ko Papakura	*Waving silently is Papakura*
Kei waenganui ko Te Puke a Ruahine	*And in-between is Te Puke a Ruahine*
Noho tonu rā ko Purapurawhetū	*Settling there also is Purapurawhetū*
Ko Ngāraratuatara	*And Ngāraratuatara*
Kia tīkarokaro atu rā	*Whilst digging my feet into the ground*
E tū ana au i te take o Pohaturoa	*I stand before Pohaturoa*
Ka rongo ia Te Kūmete	*Hearing that Te Kūmete*
Ko te nohanga tērā o Tuohonoa	*Settled at Tuohonoa*
Kia tipi atu au ki te tūranga o te	*I then creep over to where Tukutuku*
Tukutuku	*Lived, drawing to the cave of*
Me whakatata ki te ana o Tukiterangi	*Tukiterangi*

GEOTHERMAL SYSTEMS

Geothermal areas are commonly close to the edges of tectonic plates, and New Zealand's location on an active plate boundary (between the Indo-Australian and Pacific Plates) means it is prone to earthquakes and volcanoes. The country's world-class geothermal energy resource is a consequence of this activity.

Geothermal systems are found throughout New Zealand. High-temperature geothermal fields are principally located in the Taupō Volcanic Zone, which extends from Whakaari in the Bay of Plenty, southwest to Mount Ruapehu.

CREATING GEOTHERMAL SYSTEMS

The earth's interior has large quantities of heat stored in rocks and magma. At a depth of 2–3 kilometres, temperatures range from 270–350°C. The magma beneath the Taupō Volcanic Zone provides a huge heat source that has created and sustained geothermal systems for thousands of years.

GEOTHERMAL FEATURES

A geothermal system is created when rock is heated to temperatures higher than surrounding areas. When cool ground water from rain, snow melt, rivers or lakes comes into contact with the hot rocks underground, geothermal features are created.

As the water heats, geothermal liquid rises to the surface, feeding hot lakes or pools; streams carrying mineralised fluids; deposits from the mineralised waters, such as sinter terraces; steaming ground; fumaroles; and mud pools. Unique plants, animals and micro-organisms flourish in these environments.

Calderas

A caldera is a large, collapsed volcanic vent that creates a bowl shape. Calderas often contain crater lakes, such as Lake Rotorua and Lake Taupō.

Geysers

Geysers are a rare class of boiling spring. They occur when underground water in a large reservoir cannot discharge freely but is forced through a narrow opening in the earth's surface. The water pressure builds up, and the superheated mixture of water and steam intermittently erupts. Some geysers, such as Pohutu Geyser and Tohu Geyser (also called the Prince of Wales Feathers) at Te Puia, spout regularly.

Boiling springs

Boiling alkaline chloride springs occur where geothermal fluids rise quickly to the surface without being modified by contact with rocks, soil or ground water. They often occur along a line of weakness such as a fault. The cooking pools at Whakarewarewa are an example of this. They are far too hot for bathing.

Hot springs

Hot springs are natural springs with water that is warmer than normal body temperature. Springs can exist in small or large pools, lakes or flowing streams. Rotorua, especially Whakarewarewa, is acknowledged as one of six major hot spring regions of the world. Others include Yellowstone National Park in the United States, and sites in Chile, Iceland, Japan and Kamchatka (in the northeast of Russia).

Mud pools and fumaroles

Mud pools are created where there is limited hot water but abundant steam and rock material capable of breaking down into mud. Hydrogen sulphide gas (which generates the distinctive, rotten-egg smell common to thermal areas) in the steam reacts with oxygen to form sulphuric acid. This dissolves the surrounding rock, making fine particles of silica and clay. The mixture, along with the small amount of water, forms the seething and bubbling mud pools found throughout the natural thermal wonderlands of Whakarewarewa.

A fumarole is an opening or vent in the earth through which hot gases or steam escape to reach the surface.

Silica terraces

Silica terraces are formed over hundreds or thousands of years as hot geothermal waters flow out over the land and rocks, cooling and releasing silica and other minerals. New Zealand has been home to some of the largest silica terraces in the world, including the famous Pink and White Terraces, destroyed by the volcanic eruption of Mount Tarawera in 1886.

ME TĪMATA, PART 6

Ka titiro whakararo ki Te Anarata	Looking down I see Te Anarata
Te pātaka o Te Whakangehe	The storehouse of Te Whakangehe
Kia tūpato kei pahū a Waikite	Being careful that Waikite might blow
Aue rā ko te Rangianiwaniwa	Alas it is Rangianiwaniwa
Te mahuetanga o ngā puna tawhito	The last surviving mineral pool
I waihanga iho rā e Tarawera	Untouched by Tarawera

A GUIDING LEGACY

HE HUNGA ARAHI, THE GUIDES

Since the beginning of the twentieth century, an extraordinary group of people have shown the wonders of Te Whakarewarewa Valley to multitudes of tourists. With each generation of guides trained by their experienced elders, they are famed for their expert knowledge, the impeccable presentation of their culture to visitors and their personal charm.

Many of the guides at Te Puia and Whakarewarewa today are the sons, daughters, grandchildren and great-grandchildren of the famous guides who first settled the area. Guides can recite their whakapapa (genealogy), going back over twenty-five generations to the original inhabitants. They not only retell the traditional tales of past generations, they also often share their own stories with visitors.

This photo is of twin guides Georgina Rauoriwa and Eileen Rangiriri.

GUIDE SOPHIA

Te Paea Hinerangi (1832–1911), famously known as Guide Sophia, took tourists across Lake Rotomahana on the boat trip to the Pink and White Terraces prior to the eruption of Mount Tarawera.

Guide Sophia was of Ngāti Ruanui (Taranaki) and Scottish descent. She and her husband lived at Te Wairoa before the 1886 eruption, relocating to Whakarewarewa after the disaster, along with others who survived. Guiding a tour group just prior to the eruption, she witnessed a phantom waka on Lake Tarawera. The sight was interpreted as a warning of impending doom.

Sophia Hinerangi. Ref: 1/2-061777-F. Alexander Turnbull Library, Wellington, New Zealand. http://natlib. govt.nz/records/22881513

On the night of the eruption, she saved the lives of dozens of tourists and people at Te Wairoa, sheltering more than sixty people in her own humble dwelling. Sophia was appointed chief guide of the reserve in 1896.

GUIDE MAGGIE

Two women who greatly influenced the development of tourism in New Zealand were Mākereti (Maggie) Papakura and her daughter-in-law, Rangitiaria Dennan, the famous Guide Rangi. They guided many travellers through the natural wonders of the region as well as explaining the intricacies of their culture.

Mākereti Papakura (1873–1930) was born Margaret Pattison Thom to an Englishman and a highly respected Te Arawa woman, Pia Ngarotu Te Rihi. Guide Maggie, as she became known, was a beautiful, educated and much-celebrated Whakarewarewa guide, who achieved international recognition and a measure of international celebrity. In the early 1900s, she wrote a book, *Guide to the Hot Lakes District*, and formed a choir and a concert party with her half-sister Bella, who was also a guide. They travelled to Sydney in 1910 and then to London for the Empire Celebrations the following year. In 1912, Guide Maggie moved permanently to England where she married an Englishman and studied anthropology at Oxford University. Sadly, she died just before her thesis was due for examination, though it was eventually published eight years after her death. *The Old-Time Māori* is an account of the customs of Te Arawa from a woman's point of view, and has a unique place as the first extensive, published ethnographic work by a Māori scholar.

GUIDE RANGI

For more than forty years, Rangitiaria Dennan (1897-1970), was a Whakarewarewa tourist guide. Rangitiaria was the granddaughter of the great carver Tene Waitere, of Ngāti Tarāwhai, who was one of the first tutors at the national carving school, now based at the New Zealand Māori Arts and Crafts Institute, Te Puia. She married the only son of Mākereti Papakura.

Guide Rangi escorted many dignitaries, including royalty and heads of state, as well as thousands of tourists and domestic visitors to Whakarewarewa. In 1943, Guide Rangi made world headlines by hosting Eleanor Roosevelt, wife of US president Franklin D Roosevelt. During the 1953-54 royal tour, she caused a stir when she breached protocol by offering a supportive arm to the young Queen Elizabeth on a difficult section of track in the Whakarewarewa reserve.

GUIDE BUBBLES

Affectionately known as Guide Bubbles, Dorothy Huhana (1919-2006) was a well-loved guide who was born at Whakarewarewa and raised there by her grandparents following her mother's death. Guide Bubbles became an apprentice guide in 1936 and was officially registered two years later after being mentored by famous guides such as Bella Papakura. Like her predecessors, Guide Bubbles hosted many dignitaries, including Queen Elizabeth, who shared a cup of tea with her in her own home at Whakarewarewa. Guide Bubbles went on to tutor younger guides at Whakarewarewa in the 1970s. She was highly respected both locally and internationally.

ME TĪMATA, PART 7

Ka tumeke tonu au ki Te Kiri, ki Te Mimi	Shocked to see Te Kiri, Te Mimi,
Ki Te Wai o Parewhārangi	Te Wai o Parewhārangi
Kia whakatā mai au ki	I now rest at Rotokānapanapa
Rotokānapanapa	So that my face can be blown by
Kia pūhia te kanohi i te hau	the wind
Me hoki au ki te pā o Te Puia	I return back to the pā of Te Puia
Ka whakarongo āku taringa	Then turning my ears to listen to the
Ki te tangi o Rotowhio	Whistling of Rotowhio
Te okiokinga o te tini me te mano	Once a resting place for the dead
Tū ana mai ko Te Aronui a Rua	Moving I stop at Te Aronui a Rua
Takoto iho rā ko Te Arawa	There sits Te Arawa (war canoe)
Hei tohu mō te rau tau o Rotorua	Made for the centennial of Rotorua
Te moana tērā i kauhia iho rā e	The lake that was conquered by
Hinemoa	Hinemoa
Te uri o te ara kāwai o Tūhourangi	Who is of high pedigree of
E kokoia e ara e!	Tūhourangi

IMPACT OF GEOTHERMAL DEVELOPMENTS

Natural geothermal features can be easily and irreparably damaged by geothermal development. Of five major New Zealand geyser fields in existence a century ago (Rotomahana, Whakarewarewa, Orākei Korako, Wairākei and Tauhara), only Whakarewarewa still has a significant number of active geysers. The Orākei Korako field was largely drowned when the Waikato River was dammed for hydro-electricity. Development of the Wairākei field for geothermal energy resulted in a decline of ground water levels at both the Wairākei and nearby Tauhara fields and the loss of geysers and alkaline springs. Some land at Wairākei has subsided by 15 metres. Of more than 200 geysers active in the central North Island in the 1950s, fewer than a quarter remain.

ROTORUA

A marked decline of geyser activity in the Rotorua field led to a ban being placed on geothermal aquifer extraction within 1.5 kilometres of the geysers at Whakarewarewa. Restrictions were put in place to encourage more efficient use. These are set out in the Rotorua Geothermal Management Plan.

WAIRĀKEI

Just north of Taupō is the Wairākei Power Station. Built in 1958, the station was New Zealand's first geothermal field to produce electricity and, when it opened, only the second geothermal power station in the world. While the Crown owns the geothermal resource, the power station is privately owned and operated. Steam is used to drive the turbines, producing 1550 gigawatt hours of electricity annually - enough power to supply the cities of Rotorua, Taupō, Napier and Hamilton.

THE FUTURE OF GEOTHERMAL RESOURCES

The future of geothermal resources is hugely important for Māori – for the future survival of traditional Māori culture and for the tourist industry in this unique natural area.

In the Taupō Volcanic Zone, geothermal power continues to be developed by different groups. Testing throughout the region continues, to determine whether geothermal power is a possibility for the future. However, scientists have warned that the geothermal resource is fragile. For example, the drilling of private bores for domestic heating has seen the demise of many of Rotorua's geysers over the past thirty years. Extreme care needs to be taken to conserve these precious resources for future generations.

Nevertheless, the region's natural features offer significant possibilities for Māori developments. Central North Island iwi are tapping into these geothermal systems, providing some power to the national grid and also using power for their own business developments. In the past, this has come at a cost, but as technology improves, there is less destruction of the landscape.

To the northwest of Taupō is the Mōkai geothermal field, which is being developed by the Tuaropaki Trust, in partnership with Mighty River Power.

Tuaropaki Trust provides both industry and the national grid with geothermal electricity, as well as generating power for use in its milk powder plant, glasshouses, water pumps and other Tuaropaki enterprises.

Tuaropaki Trust has stressed the importance of managing this geothermal resource in a sustainable and responsible manner. All of the geothermal fluids extracted are returned underground, after their heat energy has been extracted. Methods like this will help to ensure the ongoing sustainability of this precious resource.

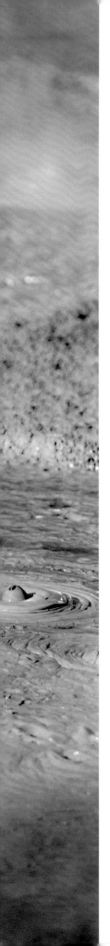

GLOSSARY

ahi māori – natural fire

ahi o te atua – fire of the gods

ahi tipua – volcanic or supernatural fire

haka – posture dance

hapū – sub-tribe

iwi – tribe

kaitiaki – guardian

karakia – prayer

Me Tīmata – I begin

Ngā Pūmanawa e Waru o Te Arawa – the eight beating
    hearts of Te Arawa

ngā waiariki – thermal springs

Otukapuarangi – the Pink Terraces

pā – fortified area of a village

Papatūānuku – the earth mother

pātere – song poem

Ranginui – the sky father

rohe – region

taonga – treasure

Te Ika a Māui – The Fish of Māui, the North Island of
    New Zealand

Te Puia – the volcano, geyser or hot spring

Te Rā – the sun

Te Tarata – the White Terraces

tohunga – expert

tūpuna – ancestor

waka – canoe

Whakaari – White Island

whakataukī – proverb

BIBLIOGRAPHY

Conly, G. 1985. *Tarawera: The Destruction of The Pink and White Terraces*. Wellington: Grantham House.

Hall, S. 2009. *Rotorua: Stories Behind the Scenery*. Auckland: Hachette.

Houghton, B. and Scott, B. 2008. *Geyserland: A Guide to the Volcanoes and Geothermal Areas of Rotorua*. The Geological Society of New Zealand.

Kingi, M. 1995. *A Preliminary Report on the History of Whakarewarewa, Draft 3 December 1995*. Rotorua: Whakarewarewa Forest Trust Research Team.

Kingi, M. Te Patere – Me Timata. Composed by Mauriora Kingi.

Lynne, B. 2003. *The Geothermal Guide to Wai-O-Tapu Thermal Wonderland*. Publishing Press.

Reed, A. W. 2006. *Legends of Rotorua*. First published in 1958. Auckland: Reed.

Royal, Te A. C. 'Hawaiki', *Te Ara – the Encyclopedia of New Zealand,* updated 9 November 2012, http://www.TeAra.govt.nz/en/hawaiki.

Stafford, D. M. 2002. *Landmarks of Te Arawa Volume 1: Rotorua*. First published 1994. Auckland: Reed.

Stokes, E. 1991. *Wairakei Geothermal Area (Some Historical Perspectives)*. Hamilton: University of Waikato.

OTHER SOURCES

www.nzgeothermal.org.nz/surface_effects

www.teara.govt.nz/en/geothermal-energy

First published in 2015 by

New Zealand Māori Arts and Crafts Institute, Rotorua, www.tepuia.com

and Huia Publishers

39 Pipitea Street, PO Box 12-280

Wellington, Aotearoa New Zealand

www.huia.co.nz

ISBN 978-1-77550-193-0

Contributing writers: Vanessa Bidois, Cherie Taylor and Robyn Bargh